URANUS AND NEPTUNE

WORLDS BEYOND

RON MILLER

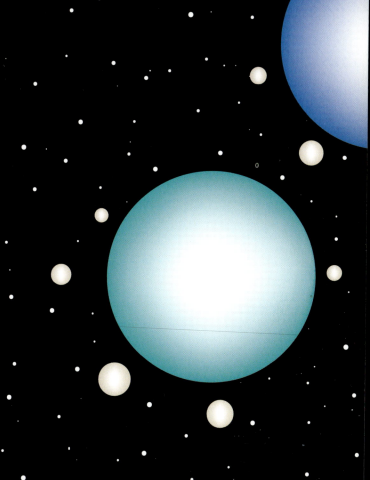

TWENTY-FIRST CENTURY BOOKS   BROOKFIELD, CONNECTICUT

This book is for Olivia Duvall.

Illustrations by Ron Miller. Photographs courtesy of NASA.

Library of Congress Cataloging-in-Publication Data
Miller, Ron, 1947–
Uranus and Neptune / Ron Miller.
p. cm. — (Worlds beyond)
Summary: Chronicles the origin and discovery of Uranus, Neptune, and their moons and discusses explorations and composition of these distant gas giants.
Includes bibliographical references and index.
ISBN 0-7613-2357-0 (lib. bdg.)
1. Uranus (Planet)—Juvenile literature. 2. Neptune (Planet)—Juvenile literature.
[1. Uranus (Planet) 2. Neptune (Planet)] I. Title.
QB681 .M55 2003   523.47—dc21   2001008480

Published by Twenty-First Century Books
A Division of The Millbrook Press, Inc.
2 Old New Milford Road
Brookfield, Connecticut 06804
www.millbrookpress.com

Illustrations and text copyright © 2003 by Ron Miller
All rights reserved
Manufactured in China
5 4 3 2 1

CONTENTS

Chapter One
The Mystery Twins 5

Chapter Two
The Story of the Solar System 11

Chapter Three
A Detective Story 16

Chapter Four
Two Strange New Worlds 22

Chapter Five
By Spaceship to Uranus and Neptune 31

Chapter Six
The Last Giant 38

Chapter Seven
Many Moons 45

Chapter Eight
Exploring the Solar System 71

Glossary 74
For More Information 76
Index 78

Astronomical symbols for Uranus

and Neptune

Neptune as photographed by *Voyager 2* [NASA/JPL]

Uranus as photographed by *Voyager 2* [NASA/JPL]

CHAPTER ONE
THE MYSTERY TWINS

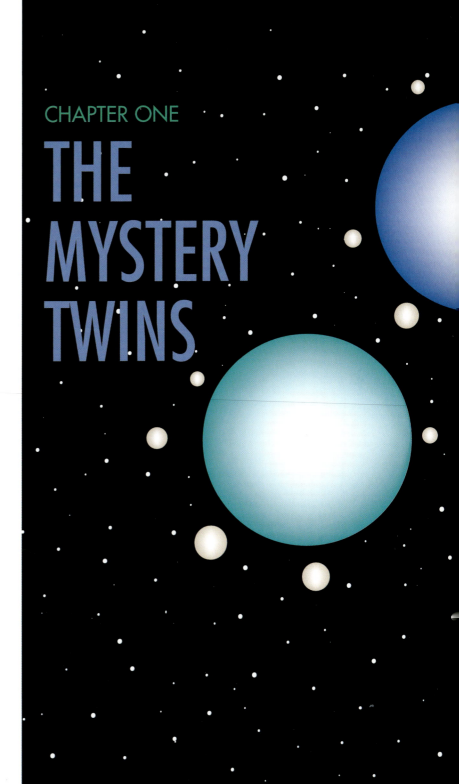

Uranus and Neptune are among the most mysterious worlds in the solar system. Only Pluto is farther from the Sun, but it is probably just a small, frozen ball of ice about the same size as Earth's Moon. Uranus and Neptune are much more unusual than that. The first worlds to be discovered by modern science, they are enormous balls of frigid, poisonous gases, nearly five times larger than Earth. They are each surrounded by a swarm of **moons**—many of which are among the most extraordinary places in our solar system.

For the first hundred years after the discovery of Uranus and Neptune, little was known other than what could be observed and measured from Earth—and that was little enough. Uranus and Neptune are very far away. Uranus is 1,690,000,000 miles (2,719,210,000 kilometers) from Earth, and Neptune is 2,700,000,000 miles (4,344,300,000 km) away—more than 7,000 and 11,250 times farther away from Earth than our Moon is. Even through the most powerful telescopes, the two **planets** look only like tiny beads, one blue, and one blue-green. This is much too small to see any details on their surfaces.

It was not until the first spacecraft, *Voyager 2*, visited Uranus and Neptune in the 1980s that astronomers learned just how very strange and mysterious they really are. But even the story of their original discovery tells of a mystery solved by much detective work.

The Discovery of the Solar System

Astronomy is the oldest of the sciences. The movement of the Sun, Moon, and stars marked the passage of days and **seasons**. It was vitally important to ancient people to know when to plant crops and when to harvest them, when the best time might be to bear children, when to expect seasons of rains, droughts, and floods, when winter would end and spring would begin. They knew when the Moon would go through each of its phases, which led to the invention of the first calendars. The crops of the ancient Egyptians depended on the annual flooding of the Nile. They looked to the sky for the rising and setting of the stars that marked the coming of the flood seasons. Ancient people also believed that the stars influenced the affairs of people, so they studied their positions carefully in order to determine what future was written in the sky. Weddings, wars, and other important events were always planned according to the best position of the stars.

So important were the movements of the stars that it was quickly noticed if anything strange or unusual occurred among them. One of the first observations was that while stars differed in brightness and color, they were all fixed motionless in the heavens. The same stars always appeared in the same relationship to

ASTROLOGY

Many people still believe that the stars and planets influence their daily lives. This belief is called **astrology**. While astrology is the ancestor of astronomy, it is not a science like astronomy. It is a **pseudoscience**, or false science. There have been many hundreds of studies made in an effort to discover a relationship between the date of one's birth and characteristics such as personality or career. No relationship has ever been found. Many newspapers and magazines publish horoscopes that pretend to give advice based on the stars, but these must never be taken seriously.

one another, night after night, with perfect predictability. All except for five. These strange stars changed position from night to night. The difference was very slight because they moved very slowly—it took weeks or even months to notice the movement—but they *did* move. The Greeks called these moving stars wanderers, which in Greek is the word *planetes*.

The five planets were named after Greek and Roman gods. The brightest of them was named for the goddess of love and beauty, Venus. The red one was named for the god of war, Mars. The second brightest was named for the chief of all gods, Jupiter, while the one that appeared to move most swiftly through the sky was named for the messenger of the gods, Mercury. Finally, the slowest moving of them all was named for Saturn, the god of time.

As interesting as the planets were, they were still not considered to be anything particularly important. The only characteristic that set them apart from the thousands of stars in the sky was that they moved. No one ever considered the possibility that they might in fact be other worlds. Not until the year 1610, that is.

The Italian scientist Galileo Galilei had been experimenting with an amazing new optical instrument that had been recently invented in the Netherlands. It consisted of nothing more than a pair of ordinary glass lenses set at either end of a wooden tube, but it had the remarkable property of making distant objects appear closer. The Dutch were immediately aware of its potential use to navigators and the military. But then Galileo did something with the telescope that no one before had thought of doing: He turned it toward the night sky.

Galileo Galilei

EXPERIMENT: GALILEO'S TELESCOPE

Needed: Two magnifying glasses

The first telescope was probably invented by accident when a curious lensmaker held up two lenses, one in front of the other, and looked through them. He must have been extremely surprised to see distant objects suddenly magnified. The first telescopes were little more than two magnifying lenses mounted at either end of a cardboard or wooden tube. Galileo's first telescope was not very powerful—it magnified only eight times, less than an ordinary pair of modern binoculars.

You can duplicate the finding of the early inventors by making your own telescope from a pair of magnifying glasses. Any magnifying glasses will do. Take one in your right hand and hold it up to one of your eyes. Take the other glass in your left and hold it in front of the first one. Now slowly move the magnifying glasses closer and farther apart until you see a sharp image of some distant object. When you do, it will look much larger than it did to your unaided eye. What else do you notice? The image is upside down. All telescopes **invert** their image this way.

This is confusing when you are looking at things on Earth, so telescopes and binoculars made for this use have an extra lens inside that turns the image right side up. But since every extra lens cuts down the brightness of the light very slightly, astronomical telescopes don't have them. It doesn't matter in the slightest to astronomers whether or not things look right side up or upside down. Most maps of the Moon, for instance, have south at the top and north at the bottom, to match the view seen in telescopes.

No one had ever thought of doing this. What was the point? The stars were nothing more than bright pinpoints of light, and everyone "knew" that the Moon was a polished, gleaming sphere of absolute purity (the dark markings were thought to be the reflection of the impure Earth). What could possibly be gained from looking at them more closely?

What Galileo learned changed not only how we look at the universe around us, but also how we look at the Earth below us. He saw that the Moon was *not* made of some pure celestial substance, but was covered with mountains and pockmarked with craters. It was, he said, "not smooth, uniform, and precisely spherical as a great number of philosophers believe it (and the other heavenly bodies) to be, but is uneven, rough, and full of cavities and prominences, being not unlike the face of the earth."

The planets, he found, were not just a special class of star but were in fact worlds, perhaps very much like our own. They were spherical, like Earth, and some of them had vague markings that might—Galileo thought—be continents and seas. Even more astonishing was his discovery that Jupiter was not only a world in its own right but that it had moons as well—four of them (though now we know that it possesses more than a dozen)—like a miniature solar system. These four moons—Ganymede, Io, Europa, and Callisto—have since been called the "Galilean" **satellites**.

Galileo's findings were censured by the church, but it was a time of great discoveries. Whole new worlds were being discovered on Earth itself, so why not in the sky as well? It did not take long for the word to get out. When Galileo's findings became widely known, people wondered: Are these worlds like our own? Does life exist on them? Are there people living there? Several books were soon published speculating about what sort of life might exist on the planets.

One such book, written by Bernard de Fontenelle in 1686, became a best-seller. *Discourses on the Plurality of Worlds* suggested that every planet was inhabited, though not at all necessarily by life resembling that on Earth. Although he knew very little more about the possible conditions on the planets other than their sizes and their distance from the Sun, he did the best he could. The people of Jupiter, he suggested, hardly ever get to know one another since the planet is so large, while, on the other hand, Mercury is so small that everyone knows everyone else. Saturn, the most distant planet from the Sun (or so it was thought at the time), was so cold that if any of its creatures were to visit Earth, they would die of the heat. Because of the cold on their planet, Saturnians live and die without ever moving from the place they were born.

The great German astronomer Johannes Kepler wrote what may be the first science-fiction novel, *Somnium*, which was published in 1634, a few years after his death. Being better acquainted with science than de Fontenelle, his description of the Moon and the sort of creatures that might live there was quite accurate, at least for the time in which he was writing. The Moon would be a very alien world, he told his readers. Nights would be fifteen days long "and dreadful with uninterrupted shadow." The cold at night

The inhabitants of some of the other worlds of our solar system as they were imagined in a book published in 1741 by Ludwig Holberg

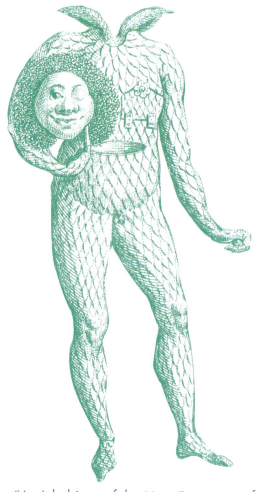

"An Inhabitant of the Moon" was one of the many strange beings encountered by Baron Münchhausen in his fictional adventures, published in 1787 by Rodolf Erich Raspe.

Inhabitants of the Moon were imagined to look like this in 1835. This illustration was inspired by the "Moon Hoax," written by journalist Richard Adams Locke. It fooled thousands of people into thinking that the famous astronomer Sir William Herschel had discovered living creatures on the Moon.

would be more intense than anything experienced on Earth, while the heat of day would be terrific. The animals that live on the Moon would have adapted to these harsh conditions. Some would go into hibernation while others would have evolved hard shells and other protection. Books like these—both fanciful and realistic— helped convince their readers that these other worlds really did exist, and that it might very well be possible that there is life on them.

CHAPTER TWO
THE STORY OF THE SOLAR SYSTEM

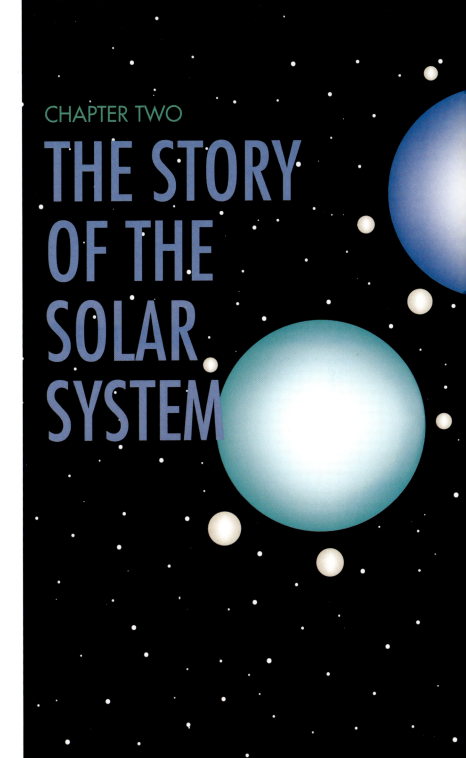

The currently accepted theory of the evolution of the solar system explains that the Sun and planets formed about 4.5 billion years ago from an enormous cloud of dust and gas. This can happen only if the cloud is large enough for the gravitation of its individual particles to start the cloud contracting and to keep the contraction going, a process called **gravitational contraction**. But once this process has begun, the cloud will shrink to a millionth of its original size very quickly. During this collapse it becomes what is known as a **protostar**.

As the center of the cloud becomes denser, its gravity increases, which in turn causes it to collapse even further. Soon, the compression of the matter at the core causes it to begin to heat up, and the center of the dark cloud begins to glow a dull red. Eventually the combination of heat and compression causes a nuclear reaction to start—perhaps only a few thousand years after the cloud first began to condense—and as soon as this happens, the protostar becomes a **star**. The increased amount of heat this produces creates an outward pressure that stops any further collapse of the dust cloud. The Hubble Space Telescope has observed young solar systems in just this phase of development. Called **protoplanetary disks**, they look like dark, bun-shaped disks, often with a dimly glowing center.

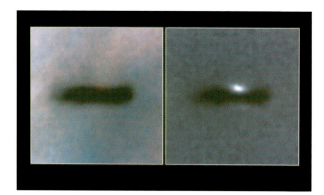

A protoplanetary disk in the Orion Nebula as photographed by the Hubble Space Telescope. In the right image the disk was photographed through filters that allow us to see glowing nebulosities above and below. This reveals the existence of the central star, which is normally hidden from us by the dark dust (left). [NASA]

Facing page: The small, rocky terrestrial planets of our solar system compared in size with the gas giants (Pluto is neither a gas giant nor a terrestrial planet.)

Within the cloud, tiny particles of dust have been colliding and sticking together, forming tiny clumps of material called **planetesimals**. As these clumps grow in size, they attract more particles. This process is called **accretion**. Most of these early collisions are relatively gentle, so the planetesimals don't knock themselves into pieces. Soon grains of dust grow to the size of rocks, then boulders, and then asteroids miles across. The whole process of growing from the size of a large pinhead to a mountain may take only 100,000 years or so. By this point the process begins to slow down—most of the original dust and gas is being used up and the cloud is growing thin. Several stars have been observed with large, thin disks of dust surrounding them—such as Beta Pictoris—which may be solar systems in this same stage of development.

As the planetesimals grow larger, they begin to move faster and the collisions between them become more violent. Now instead of accreting, some of them shatter into pieces. The increasing size and gravity of the planetesimals is the cause of the higher speeds. The few planetesimals that are large enough to survive the collisions grow even larger, devouring the debris from the unluckier smaller objects. Once the process of accretion begins, it grows very quickly. Earth may have gone from a cloud of dust to a body nearly its present size in as few as 40 million years.

Two Kinds of Planets

There are two types of planets in our solar system: **gas giants**, which are made mostly of light elements such as **hydrogen** and helium, and **terrestrial planets**, which are made mostly of rock

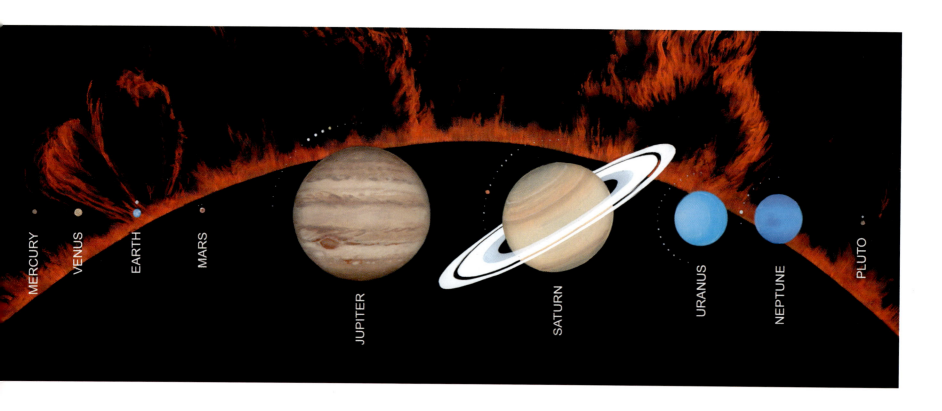

A family portrait: the planets and moons of our solar system shown to the same scale as the Sun

and metal. Earth is a terrestrial planet (and it gave its name to the category, since "terrestrial" comes from the Latin word for "earth"), as are Mars, Venus, and Mercury. Our solar system contains four gas giants: Jupiter, Saturn, Uranus, and Neptune. The four terrestrial planets lie close to the Sun while the four gas giants lie much farther away. Pluto, the most distant planet, might seem to be an exception to this, but scientists believe that it might not be a true planet. Instead, it may be an icy body captured from the vast cloud of similar objects that circle the Sun far beyond the present orbit of Pluto.

This division into two groups of dissimilar planets occurred during the formation of our solar system. It was determined mainly by the temperatures involved. In the cold outer regions of the original gas cloud, hydrogen-rich ices, such as water ice, easily formed. But in the hot inner region, only metals and silicates could exist. Lighter elements, such as hydrogen and helium, were unable to become solid. This is why the planets closest to the Sun, such as Mercury, Venus, Earth, and Mars, are made mostly of rock and metal, while the larger, cold planets that are farther away are made of lighter materials, such as ice and gas. Studying the outer planets is like studying fossils of the original gas cloud from which the solar system formed.

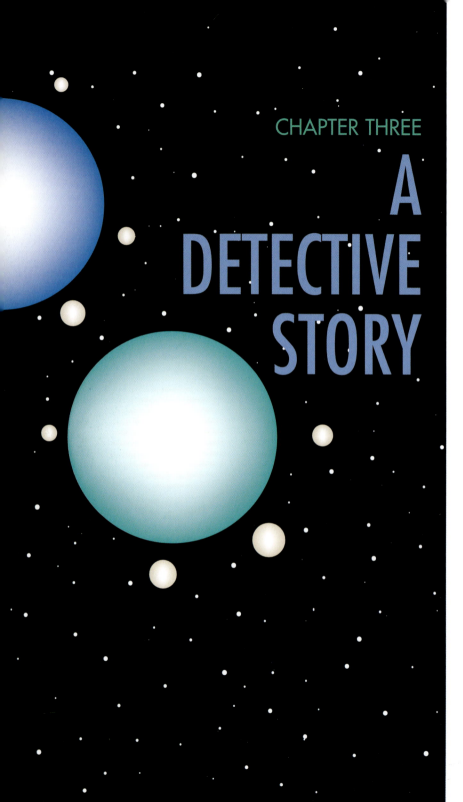

CHAPTER THREE
A DETECTIVE STORY

As telescopes became better, astronomical knowledge of the planets grew. But no discovery was quite as astonishing as that made in 1781. While doing a routine survey of the stars with a homemade telescope not much more powerful than one an amateur astronomer might own today, self-taught astronomer William Herschel discovered an entirely new planet, one that no one had even suspected existed. One reason no one had ever bothered to look for other planets was the assumption that there *couldn't* be any others. The five known planets along with the Sun and Earth made seven bodies altogether, a mystical number that seemed perfect to many people (just as some people today think the number 13 is unlucky).

At first Herschel thought he had discovered a new **comet**, but when he worked out the body's orbit he realized that it must in fact be a planet. Most comets have highly **elliptical**—oval-shaped—orbits that are very different from the nearly circular orbits of most of the planets.

Herschel's news took the world by storm, and there was a rush to name the new planet. "Georgium Sidus," or "George's Star," for King George III (which was Herschel's preference) and "Herschel" after the astronomer (which was certainly not his preference!) were two popular candidates. In fact, for many years the planet was known as Herschel, but the German mathematician Johann Bode suggested Uranus, after the Greek god of the sky, and it finally won out. The argument, which was a good one, was that all of the other planets are named after Greek and Roman gods, so why break with tradition?

It turned out, once older records were checked, that Uranus had been observed many times in the past, for nearly a century, but had not been recognized for what it was. It is so slow moving that it was mistaken for a star. (If one knows exactly where to look and if the sky is dark and clear enough, Uranus can just barely be made out by the naked eye. Astronomy magazines such as those listed at the end of this book will tell you where to look.) But when these old records were gone over carefully, astronomers realized that they revealed something even more amazing than a new planet.

Using these early observations, it was possible to calculate a very accurate orbit for Uranus. But there was a problem: Uranus didn't seem to want to agree with the calculations. It wasn't exactly in the position, either in 1781 or in the years following, that the numbers predicted. In some years it seemed to lag a little behind its expected position and in other years it seemed to move too quickly. Where was the mistake being made and how? In 1834 the Reverend T. J. Hussey of Kent, England, made a startling suggestion: Perhaps the fault wasn't in the mathematics at all. What if there was yet one more unknown planet orbiting beyond Uranus? Its gravitational pull upon the inner planet might account for everything. When the unknown planet was ahead of Uranus in its orbit, it tugged on Uranus, making it go a

Urania, the goddess of astronomy, depicted in a nineteenth-century engraving

Sir William Herschel

Because Uranus was not always exactly where it was expected to be, astronomers suspected that an unknown planet might exist. When the unknown world was behind Uranus, the pull of its gravity slowed Uranus down slightly. When it was ahead of Uranus, the pull of its gravity made Uranus speed up a tiny amount. By carefully measuring these disturbances—or perturbations—in the position of Uranus, astronomers were able to predict the location of the mysterious new planet.

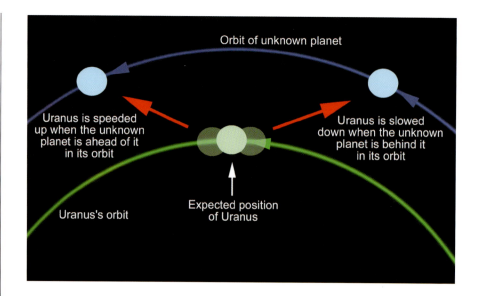

little bit faster. When it was behind Uranus in its orbit, it pulled on Uranus, making it go a little slower.

Hussey suggested that it might be possible to predict the location of this mysterious planet by working backward from its effect on Uranus. For instance, when Uranus was being slowed down, he knew that the unknown planet was behind it, and when Uranus was being sped up, he knew that the unknown planet was ahead. By carefully measuring these changes, he thought it might be possible to calculate exactly where the new world could be found.

A brilliant young math student at Cambridge University, John Couch Adams, took up the challenge and by 1843 had worked out just where he thought the new planet ought to be. He sent his results to George Airy, the Astronomer Royal. Unfortunately,

THE MUSICIAN-ASTRONOMER

Sir William Herschel was born in Germany on November 15, 1738, the fourth of ten children of Isaac Herschel, an oboist in a military band. William also became a musician. When he was 19, he emigrated to England, arriving with only a single French coin and his musical talent to support him. He managed a meager living teaching music from town to town. An accomplished organist, he was eventually appointed concert master of the city of Bath in 1776.

Meanwhile, Herschel had been reading a book on astronomy and became fascinated with the subject. He could not afford to buy a telescope so he decided to make one—moreover, he decided to make the most powerful one ever built.

In 1775 Herschel built a telescope 7 feet (2.13 m) long that was more powerful than any other telescope in the world. But even this did not satisfy him. In 1781 he managed to construct an instrument 30 feet (10 m) long. He spent every clear night observing the sky. His sister Caroline worked with him, staying up all night to record his observations and spending the following days arranging and summarizing the results of their work. It was with this telescope that Herschel not only discovered the planet Uranus, but also ensured that he would never have to work as a musician again. The fame this discovery brought Herschel gained him the post of Astronomer Royal (the official astronomer of the royal court). The handsome salary that came with it allowed him to devote all his time to astronomy.

Herschel continued to make important contributions to astronomy. Before he died in 1822, he discovered new moons of Saturn and Jupiter and compiled a vast catalog of stars. He discovered the nebulae, clouds of gas in space that look like small patches of light. When he observed the nebulae, his telescope was not powerful enough to distinguish between true gas clouds and those made up of clusters of millions of stars. These star clusters are actually galaxies, though some of them, such as the Andromeda Nebula, still retain their old name. He also studied sunspots and many other things. It can be said that modern astronomy was founded by Sir William Herschel.

The huge telescope that William Herschel built when he discovered Uranus

(19)

BODE'S RULE

A strange mathematical relationship between the orbits of the planets was discovered by the German astronomer Titius. It was popularized by his colleague, Johann Bode, in 1772, and it soon became known (somewhat unfairly) as Bode's Rule. It works like this: Bode and Titius wrote down a series of 4s. Beneath the first 4 they placed a 0. Under the second 4 they placed a 3, then 6, 12, 24, etc., doubling the number each time. They added these pairs of numbers and then divided by ten, like this:

4	4	4	4	4	4	4	4	4
0	3	6	12	24	48	96	192	384
0.4	0.7	1.0	1.6	2.8	5.2	10.0	19.6	38.8

What they noticed immediately was that six of the numbers almost exactly matched the distances from the Sun in astronomical units of the then-known planets (an astronomical unit, or **AU**, is the distance of Earth from the Sun):

	MERCURY	VENUS	EARTH	MARS	?	JUPITER	SATURN
Predicted Distance	0.4	0.7	1.0	1.6	2.8	5.2	10.0
Actual Distance	0.4	0.7	1.0	1.5	2.8	5.2	9.5

Bode's Rule got a real boost when Uranus was discovered in 1781 and it fit into the chart almost perfectly. The rule said that the distance from the Sun to the next planet beyond Saturn should be 19.6 astronomical units, and the actual distance of Uranus turned out to be 19.2. And what about that question mark in the fifth position? Many astronomers, impressed with the accuracy of Bode's Rule, began searching for the planet they thought must exist in order to fill the fifth position. In 1801 the Italian astronomer Giuseppe Piazzi discovered the first and largest **asteroid**, which he named Ceres—and it was at exactly the distance predicted by the rule! Ceres is scarcely large enough, however, to be called a planet. Since 1801, of course, thousands of other asteroids have been discovered between Mars and Jupiter.

The discovery of Neptune, however, proved to be something of a blow to the credibility of the Rule. It orbits at 30 AU and there is no place on the chart for it. The Rule does not predict another planet until 38.8 AU. Pluto, discovered in 1930, conformed somewhat better. Its average distance from the Sun is 39.4 AU.

Is Bode's Rule a law of physics? Can planets form only at these distances? Or is the rule only a happy coincidence? Most astronomers don't consider it a law, but rather only as a handy tool for remembering the distances of the planets. Still, it is curious that something seems to have compelled each planet in the solar system to form about twice as far from the Sun as the next one in. No one knows why this is. Perhaps we shall have to wait and see if Bode's Rule applies to the planets of solar systems other than ours.

for various reasons, Airy did nothing with Adams's calculations until 1846, when a young French astronomer, Urbain Le Verrier, published the results of his own calculations. Le Verrier had used the same reasoning that Adams had, and his predicted location for the new planet was almost exactly the same. As soon as Airy saw this, he immediately assigned two astronomers—James Challis and William Lassell—to search for the planet.

Challis believed he saw the new planet on August 4 and again on August 12, but had to double check his observations. It was important that his figures were correct and the planet was identified before he made any official announcement. Before he had a chance to do this, however, Johann Galle and Heinrich d'Arrest of the Berlin Observatory in Germany, using Le Verrier's figures, found and identified the planet. They announced their discovery to the world. That's when the problems began.

Adams, an Englishman, had been the first to predict the location of the new planet, while it was the Frenchman Le Verrier's mathematical work that led to its actual discovery. A huge international argument (in which, to their credit, neither Adams nor Le Verrier took part) broke out as to which country—France or England—could claim credit for the discovery of the new world. Germany, too, put in a bid, since, after all, it was a pair of German astronomers who actually found the planet. The French wanted to name it Le Verrier, to the horror of every other country, but tradition and good sense won out in the end, and the new planet was named Neptune, after the Roman god of the sea.

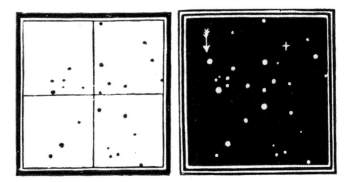

On the left is the star chart Galle used in his search for Neptune. On the right is the sky as he saw it, with Neptune's position marked with an arrow. The small cross indicates where Le Verrier predicted the planet would be.

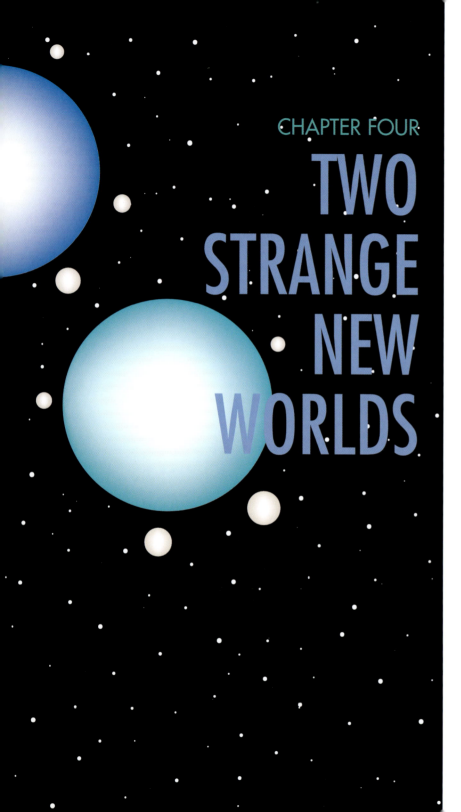

CHAPTER FOUR
TWO STRANGE NEW WORLDS

Uranus and Neptune are so far from the Sun and so difficult to observe (from Earth, Uranus is the size of a golf ball seen at a distance of 1.5 miles [2.5 km]), that very little was learned about them during the 140 years following the discovery of Neptune. Since either planet appears as only a pinhead even in the most powerful telescopes, it was almost impossible to see any details. Astronomers determine the rate of rotation of a planet by timing the movement of features on its surface. No one was able to tell for certain how long it took Uranus and Neptune to rotate since neither planet had any distinctive markings or cloud patterns.

One thing astronomers were able to determine was the planets' composition. This was accomplished by the use of an instrument called a **spectroscope**, which can detect what elements exist on other planets by studying the light reflected from them. This was first done in the nineteenth century, when scientists learned that Uranus and Neptune were similar to Jupiter and Saturn in being composed mostly of hydrogen with some helium and methane. But even the spectroscope was not completely reliable where Uranus and Neptune were concerned. Because the

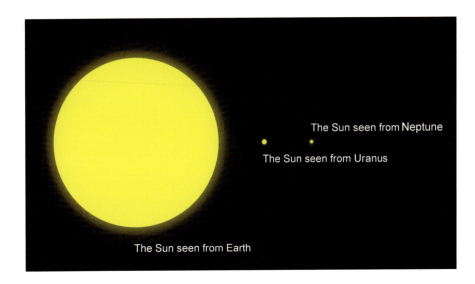

The Sun seen from Neptune
The Sun seen from Uranus
The Sun seen from Earth

This is the size of the Sun as it looks from Earth compared with how it appears from Uranus and Neptune. To get a more realistic idea of the difference, set this book up open to this page. Then step back until a dime held at arm's length just covers the large Sun. This is how large the Sun looks in Earth's sky. Now see how small the Sun looks from Uranus and Neptune!

planets are so small and dim, exact measurements with Earth-based instruments are extremely difficult to make. Scientists were sure that hydrogen, helium, and methane existed on Uranus and Neptune, but in what proportions? And what other gases or compounds were there?

What Astronomers Did Know

Using Earth-based telescopes, astronomers were able to determine the length of Uranus's and Neptune's year, the number and orbits of their moons, the sizes of the planets, and their masses.

Uranus orbits 1,782,000,000 miles (2,869,600,000 km) from the Sun—nearly twenty times farther than Earth. It takes 84 years

THE SPECTROSCOPE

In 1666 Sir Isaac Newton split sunlight into its separate colors by means of a glass prism. The sunlight was spread into a broad band like a rainbow, proving that white light was really a combination of different colors. This band of color is called a **spectrum**.

When a substance—either solid, liquid, or gas—is heated to incandescence at a high pressure, it gives off a continuous rainbow of colored light, or a spectrum. Each spectrum is unique to the element that produces it, like a fingerprint. For example, the element sodium produces two bright yellow lines in a particular position—no other element will do this. Each element and each compound of elements has its own unique color "fingerprint" by which it can be identified. By examining the spectrum of any light-emitting object, it is possible to determine the elements of which it is made.

The instrument that astronomers use to study the spectrum is called a **spectrograph**. It consists of either a glass prism or a **diffraction grating**, which is a glass plate with many thousands of fine lines. The diffraction grating splits the incoming light into its component colors so that the resulting spectrum can be observed and photographed. The colors you see on the front of a CD are actually a spectrum created by the fine grooves in the surface—when this happens, the CD is acting exactly like a diffraction grating.

---

to circle the Sun just once (only 2.6 Uranus years have passed since its discovery). It is a large planet, 31,765 miles (51,118 km) in diameter—four times that of Earth. Even though it is a large planet, its surface gravity is only 88 percent that of Earth. This is because Uranus is made mostly of very light materials while Earth is made of rock and metal.

Even less was known about Neptune before the first spacecraft reached the planets. It is so far away—2,792,400,000 miles (4,496,600,000 km), thirty times farther from the Sun than Earth—that it appears as a barely discernible greenish disk in only the largest telescopes. At that great distance it takes 164 years to make one orbit around the Sun. One full Neptune year has not yet gone by since its discovery—this won't happen until the year 2010.

Neptune is slightly smaller than Uranus—30,778 miles (49,530 km) in diameter—and until 1989 was thought to have only two moons. Like Uranus, Neptune seemed to be composed mostly of methane ice and other ices. Other than the slight difference in size, the two planets appeared to be near twins of one another.

A Surprising Discovery

Before the first spacecraft flybys of *Voyager 2*, launched in 1977, Neptune was known to have two moons and Uranus was known to have at least five. The orbits of Uranus's moons revealed a very unusual fact about the planet.

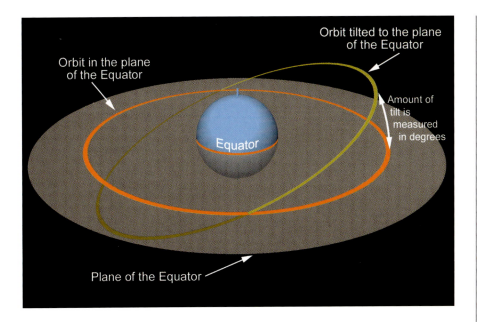

Any object orbiting directly above the equator of a planet, at whatever distance, is said to be orbiting in the **equatorial plane**. The red orbit in the diagram is in the equatorial plane of the planet. If an object's orbit is tilted in relation to the plane, as in the case of the yellow orbit, this tilt is measured in degrees. An orbit in the plane itself is tilted at 0 degrees, and an orbit that passes directly over the poles of the planet is tilted at 90 degrees.

A planet rotates about its **axis** (*plural* axes), an imaginary line that runs from pole to pole, just as the wheels of a car spin on the axle that connects them. Most of the axes of the planets in the solar system are tipped to some degree. Their equators lie more or less in the plane of the solar system, with their north and south poles pointing "up" and "down." Jupiter's axis is tipped only 3 degrees to the plane of its orbit, while Earth's is tipped 23.45 degrees. Mercury has the least tilt of any of the planets, only about a tenth of a degree.

Most of the moons in the solar system orbit their planets in the same plane as the planet's equator. If the orbits of the moons are all equally tilted then the planet must be tilted too. When astronomers looked at the tilt of the orbits of Uranus's moons, they were astonished to discover that Uranus's axial tilt had to be more than 90 degrees. This is greater than any other planet's was known to be at that time (although it is now known that Pluto has a similarly great axial tilt).

The tilt of a planet's axis causes the changing of the seasons during the course of a year. A planet like Mercury or Jupiter, which has little or no axial tilt, experiences no seasons. Since Earth's axis is tipped 23.45 degrees, the Northern Hemisphere is tipped toward the Sun during the first half of the year, then six months later, the Southern Hemisphere is tipped toward the Sun. The hemisphere that is tipped toward the Sun gets the most direct sunlight, so it is summer when that happens. When a hemisphere is tipped away from the Sun, sunlight comes in much more at an angle and it is colder, so it is winter.

Uranus, however, is tipped almost onto its side, so that its axis is almost in the same plane as its orbit. This means that for half of its year its North Pole is pointed directly at the Sun and for half of its year its South Pole is pointed directly at the Sun. Since Uranus's year is 84 Earth years, not only do the poles experience seasons 21 years long, they also experience 42-year-long periods of daylight and darkness. Imagine living in a place where more than 40 years pass between sunset and sunrise!

Why does Uranus have such an extreme tilt, unlike that of any other planet? One theory is that Uranus was hit by an asteroid as large as Earth sometime in its distant past. The collision literally

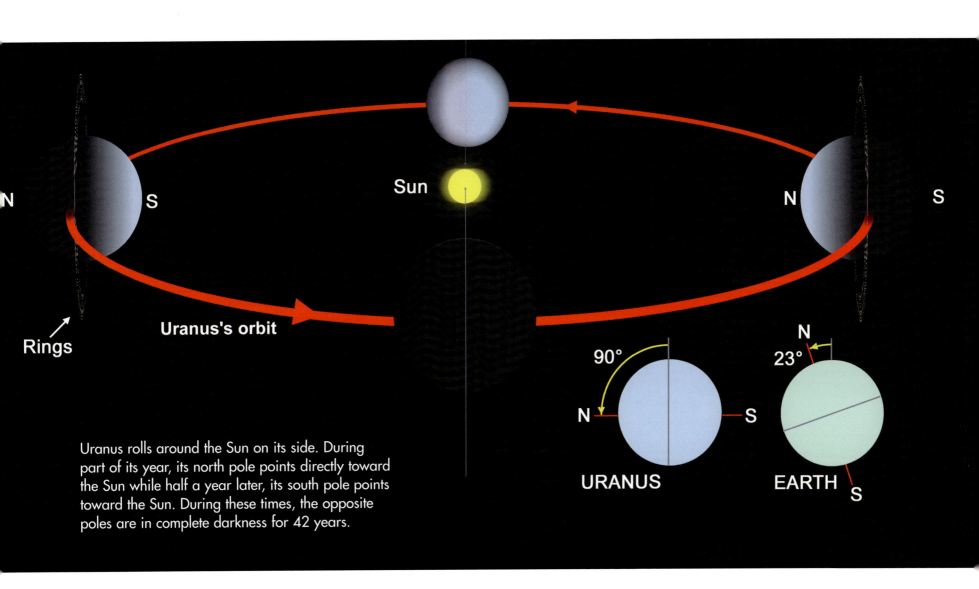

Uranus rolls around the Sun on its side. During part of its year, its north pole points directly toward the Sun while half a year later, its south pole points toward the Sun. During these times, the opposite poles are in complete darkness for 42 years.

knocked Uranus onto its side. One problem with this theory is that Uranus's moons orbit in the plane of its equator, so that whatever affected Uranus must also have affected the moons. A collision would probably only have affected the planet. Uranus's tilt is still one of the great mysteries of the solar system.

More Surprises

Another surprising discovery about Uranus was made nearly ten years before the first spacecraft arrived at the planet. Astronomers knew that Uranus was going to pass in front of a star on March 10, 1977. It is an important event when a planet does this. Measuring the time between when the star disappears and reappears can provide an accurate measurement for the size of a planet. Also, the starlight, after passing through the outer fringes of a planet's atmosphere, can be examined by a spectroscope, revealing details about the composition of the planet's atmosphere.

In order to rise far above Earth's atmosphere, which interferes with delicate observations, a group of astronomers installed a 36-inch (91.5-cm) telescope aboard a NASA jet. Flying at 41,000 feet (12,500 m), the astronomers had most of Earth's turbulent atmosphere below them. As they watched Uranus approach the star, the scientists were astonished to see the star dim, brighten, and then dim again, flickering on and off at least five times before the planet covered the star. The astronomers were mystified. What could have caused the star to do that?

The scientists watched the star carefully as it emerged from behind Uranus. The flickering repeated, this time in reverse order.

Facing page: Uranus might have gotten its strange axial tilt when it was hit by a huge asteroid millions of years ago. The impact might have knocked the planet onto its side.

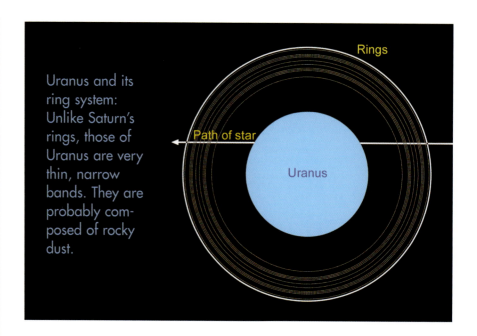

Uranus and its ring system: Unlike Saturn's rings, those of Uranus are very thin, narrow bands. They are probably composed of rocky dust.

The only thing they could imagine that could cause this was a ring—or rather a series of narrow rings—circling Uranus. They had to be very unusual rings, too: dark and very narrow, like thin bands surrounding the planet, totally unlike the broad, bright rings that surround Saturn. It was the first possible ring system to be discovered around another planet since Saturn's rings had been discovered more than 300 years earlier.

But what were the rings? What did they look like? What were they made of? It was impossible to find out from Earth; the scientists would have to wait until the first spacecraft arrived at Uranus. When this finally happened, they got more than they bargained for.

CHAPTER FIVE
BY SPACESHIP TO URANUS AND NEPTUNE

So far, the first and only spacecraft to visit Uranus and Neptune has been *Voyager 2*, which was launched in 1977. Over the course of the next 12 years *Voyager 2* made a grand tour of the outer solar system. It first encountered Jupiter, which it flew by in 1979. The swing past the giant planet gave the spacecraft a boost in speed, like a slingshot, and also changed the direction it was headed. *Voyager 2* next flew by Saturn in 1981, where it again got a boost in speed and a change in direction. It arrived at Uranus in January 1986, nine years after it had been launched from Earth. Three years later, it flew by Neptune.

The large, 1,797-pound (815-kg) spacecraft was loaded with instruments and experiments. In addition to its propulsion, power generator, and communication antennas, it carried two TV cameras, a plasma detector, a cosmic-ray detector, infrared interferometer, spectrometer, and radiometer, an ultraviolet spectrometer, a magnetometer, and other instruments. The information transmitted back to Earth was received by teams of scientists, each team dealing with the data returned by one specific instrument.

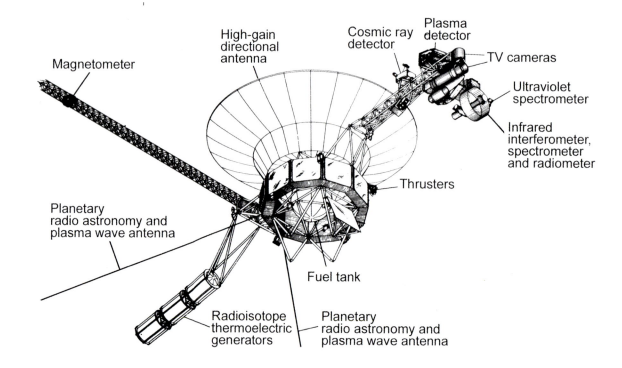

Voyager 2 spacecraft
[NASA/JPL]

At first sight, few things could have been more disappointing than the first close-up pictures of Uranus, taken by *Voyager 2* in 1986 (after a journey of nearly ten years). After the spectacular images received from Jupiter and Saturn, scientists had eagerly awaited the first images of Uranus. But instead of another world of spectacular swirling clouds, as on Jupiter and Saturn, what they got looked like a blue pool ball, nearly as featureless as a child's balloon.

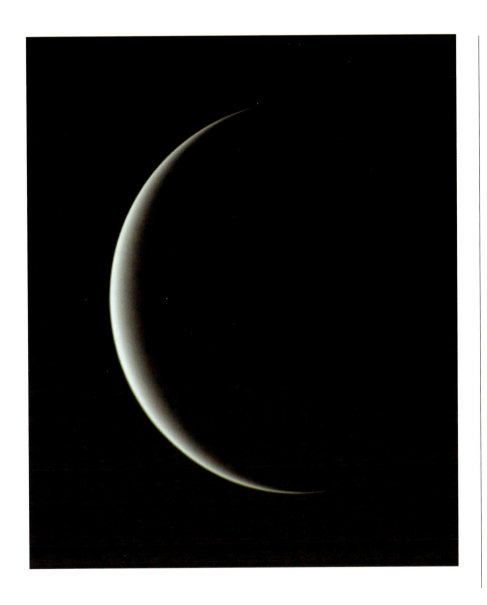

Voyager 2 caught this view of Uranus. [NASA/JPL]

FAST FACTS ABOUT URANUS

DIAMETER: 31,765 miles (51,118 km)—4 times larger than Earth

MASS: 14.5 times more than Earth

SURFACE GRAVITY (at equator): 0.88 that of Earth

LENGTH OF DAY: 17.24 Earth hours

LENGTH OF YEAR: 83.75 Earth years

DISTANCE FROM THE SUN: 1,782,000,000 miles (2,869,600,000 km)—19.16 times farther from the Sun than Earth is

---

The beautiful blue color is due in part to the deep haze that covers the planet. This haze—a smoglike layer in the upper atmosphere—scatters blue light just as oxygen **molecules** scatter blue light in the atmosphere of Earth. If you were hovering just above the cloud tops, you would probably find yourself lost in a dense, hazy mist. Another reason for the blue color is that the methane in Uranus's atmosphere absorbs red and orange light and reflects blue and green.

The lack of colorful clouds on Uranus is related to the weather. Weather on Earth is determined by energy from the Sun. Seen from Uranus, the Sun appears 19 times smaller than it does from Earth and provides 400 times less illumination and warmth. In fact, the average temperature at the top of Uranus's clouds is a frigid −360°F (−234.4°C). Uranus is so far from the Sun that there just is not enough energy being received to generate much in the way of weather. Because of the lack of weather—and the extremely long seasons caused by Uranus's axial tilt—there are none of the striking cloud patterns that make Jupiter so distinctive. All that can be seen on Uranus are wispy streaks of white and pale blue clouds.

One of the biggest thrills of the *Voyager 2* flyby was the confirmation of the rings surrounding Uranus. While they had been detected from Earth in 1977, no one had been able to see or photograph them. They were just too thin and dark. Astronomers were pleased to have the existence of the rings confirmed by the spacecraft, which photographed the rings as it flew past the planet.

The rings are nothing at all like the broad, flat, bright rings that surround Saturn, or even the thin, dusky ring around Jupiter.

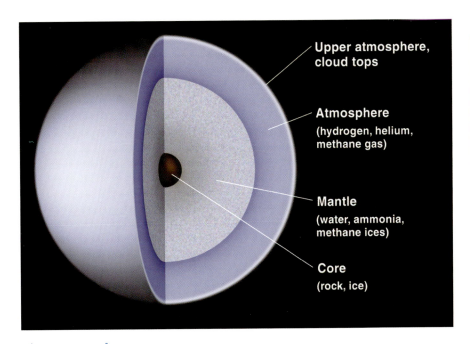

The interior of Uranus

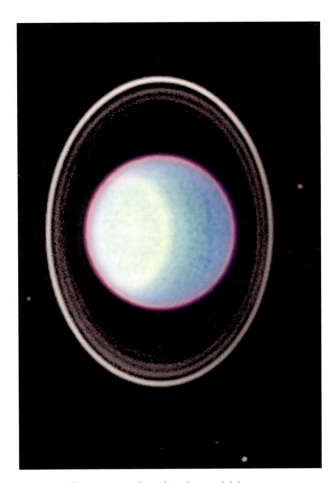

A view of Uranus taken by the Hubble Space Telescope shows both the planet and its rings. The south pole of Uranus is pointing toward the Sun. The small dots are moons. [Space Telescope Science Institute]

Instead, Uranus's rings are delicate bands that resemble concentric loops of string more than anything else. They are not solid like strings, though. Seen close-up they would look like thin, wispy bands of smoke. There are at least nine rings, each separated by a wide gap. They are very dark, resembling powdered coal more than the chunks of ice that make up Saturn's rings. Scientists suspect that they are made of dust instead of ice. The rings are probably shaped by the many moons that circle Uranus and affected by the complicated gravitational pulls from each moon.

TUG OF WAR

From Cordelia, the innermost satellite of Uranus, the giant planet looms only 15,000 miles (24,000 km) away. Cordelia, along with Ophelia, is one of the two shepherd moons that help shape the outermost ring of Uranus. The inner rings, seen edge-on, look like ghostly clouds, their shadows casting thin dark lines on the planet.

Four of the solar system's planets are known to have rings: Saturn, of course, Jupiter, Uranus, and Neptune. None of these ring systems are alike, yet they all share certain qualities.

The moons that orbit each planet have an effect on its rings. As the moons orbit, their gravity pulls on the particles that make up the rings. Sometimes the **orbital period** (the time it takes to go around the planet) of a ring particle has a simple ratio with the orbital period of a moon, such as 3:1, 2:1, or 5:2. This means that every time the particle goes around three times, the moon goes around once, and so on. Every time the moon and particle line up, the moon

gives the particle a little tug on a regular basis. Eventually, the particle will move into a new orbit. After enough time, the moon will clear an open space in the ring.

There are also "shepherd moons" that actually orbit within some ring systems. They help to clear areas and shape the rings. The extreme narrowness of Uranus's rings is probably caused by tiny moons orbiting between them. For example, the moons Cordelia and Ophelia orbit on either side of one of Uranus's most prominent rings. The effect of their gravity on the ring particles maintains the ring's shape in much the same way a pair of sheepdogs keeps a herd of sheep from wandering.

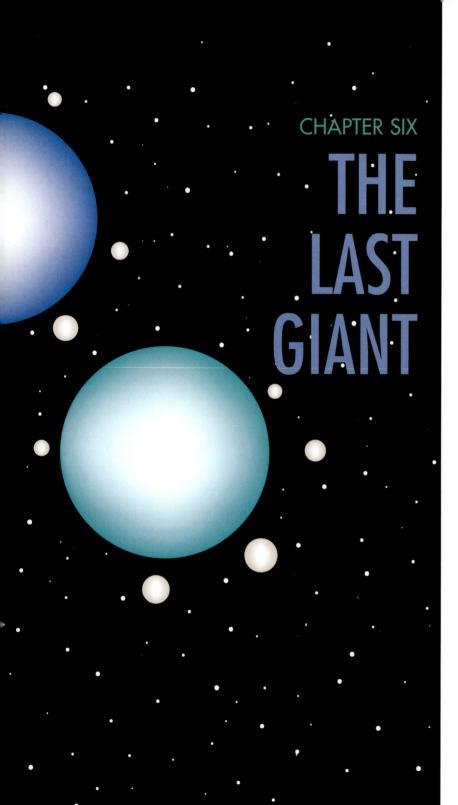

CHAPTER SIX
THE LAST GIANT

Of the four gas giant planets in our solar system, Neptune is the farthest from the Sun, at 2.8 billion miles (4.5 billion km). It takes 164 years for it to orbit the Sun just once, so Neptune has not quite made one full orbit since it was discovered in 1846. Neptune is a little smaller than Uranus, though it is still large enough that it could hold nearly sixty planets the size of Earth.

Since Neptune resembles Uranus so much, except for being a lot colder, scientists expected it to be pretty much a twin of Uranus—another featureless blue ball. But they should have known that the solar system is full of surprises. *Voyager 2* discovered that Neptune is not a replica of Uranus. Neptune has prominent bands of blue clouds, wispy white clouds, and, even more surprisingly, a huge dark-blue oval spot. This is an enormous, hurricanelike storm, similar to the Great Red Spot on Jupiter. It has been named the Great Dark Spot. It is nearly 8,000 miles (12,500 km) wide—an area as wide as the diameter of Earth. It rotates like storms on Earth, with winds blowing as fast as 700 miles an hour (1,126 km an hour)! Several smaller blue-and-white oval storm systems are scattered elsewhere on the planet.

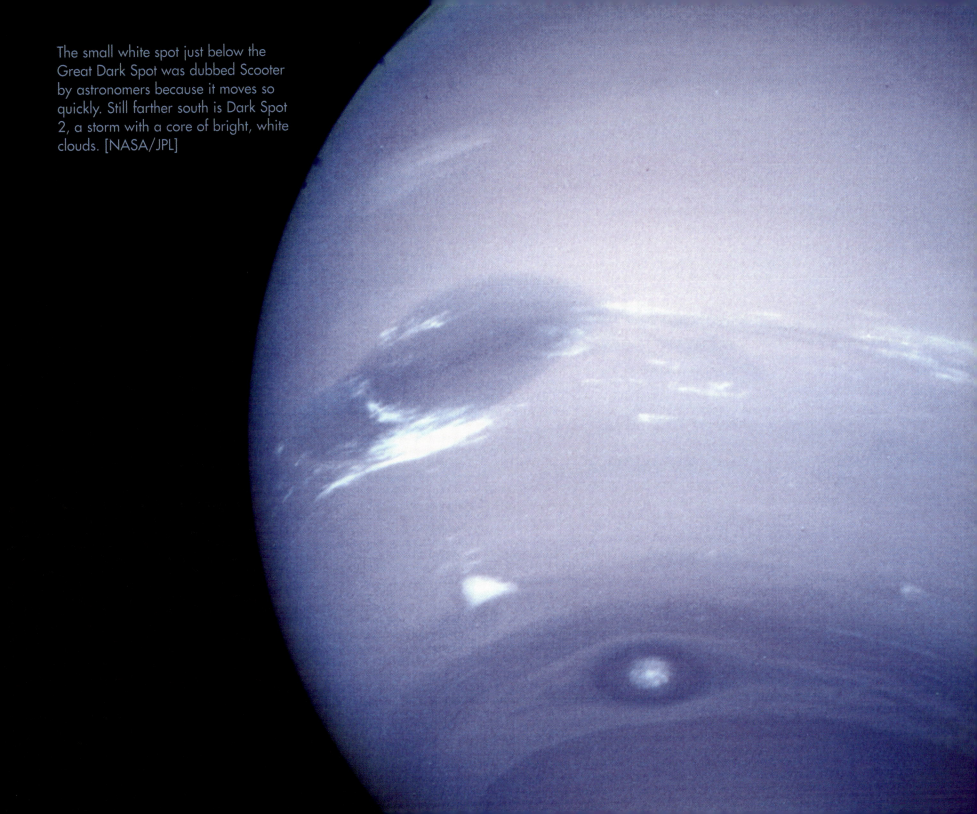

The small white spot just below the Great Dark Spot was dubbed Scooter by astronomers because it moves so quickly. Still farther south is Dark Spot 2, a storm with a core of bright, white clouds. [NASA/JPL]

The wispy white clouds are very high in the atmosphere, as high as 30 to 45 miles (50 to 75 km) above the main cloud deck. They are probably made of water ice crystals. The dark blue methane clouds—including the Great Dark Spot—are much lower. Most of the high, white clouds are stretched into long, straight bands because of the planet's rapid, 16-hour rotation.

What Powers Neptune's Storms?

Heat from the Sun provides the energy for Earth's weather. The Sun warms Earth's soil. This warmth causes the air above to circulate, like water being heated in a pan. Solar energy also warms the water in lakes and oceans. This warmth not only causes the air above to circulate, it also evaporates the water, filling the atmosphere with water vapor. It is the circulation of the air in the form of wind that carries the water vapor all over the planet in the form of clouds, snow, and rain.

But the Sun is thirty times smaller in Neptune's sky than it is in Earth's sky and provides only 1/900th of the light and heat. Where does Neptune get the energy to power its weather? Could the planet be providing its own heat from some internal source? This is how Jupiter's weather is powered. Gravitational contraction deep within the planet produces more heat than Jupiter receives from the Sun. Its weather is entirely independent of the Sun's energy. Could the same thing be happening on Neptune? Could warm currents rising from deep within Neptune's atmosphere keep its storms raging?

High, white clouds cast dark shadows on the blue clouds below them. Strong winds have stretched the clouds into long streaks. [NASA/JPL]

(41)

Voyager 2 sent back temperature readings that proved that this was indeed the case. Neptune, it showed, actually produces more heat than it receives from the Sun. In fact, Neptune has the highest ratio of internal heat outflow compared with incoming sunlight of any planet—a ratio of 2:7. That is, for every two units of heat coming from the Sun, seven are coming from inside Neptune. If the Sun were suddenly to go out, Neptune would hardly notice.

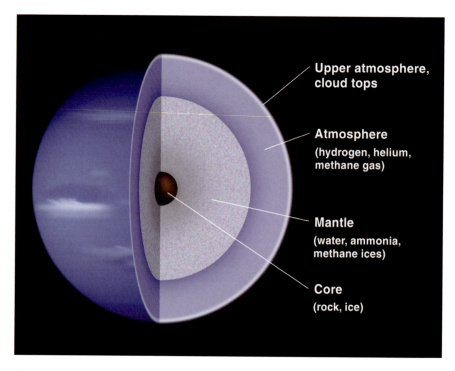

The interior of Neptune

Neptune's Atmosphere

Neptune's atmosphere, like that of Uranus, is made mostly of hydrogen (85 percent), helium (13 percent), and methane (2 percent). Just as water cycles through several stages in Earth's weather system—from water to water vapor to ice crystals—methane does the same on Neptune. When ultraviolet light from the Sun acts on methane high in the atmosphere, it breaks it down into **hydrocarbons** such as ethane, acetylene, and other particles that form a haze over the entire planet.

These haze particles, which are heavier than the original methane, sink into the lower atmosphere. It is colder there, and they freeze. As they sink even further, they reach the lower, warmer levels of the atmosphere. The heat causes the particles to evaporate back into gases. These hydrocarbon gases mix with the hydrogen in the lower atmosphere, where the heat and pressure cause the hydrogen to combine with the hydrocarbons, creating methane again. Since methane is a very light gas, it rises, eventually returning to the upper atmosphere where the whole cycle starts over again.

More Rings

Neptune was also found to have rings. As with Uranus's rings, they were first detected from Earth and then later photographed by *Voyager 2*.

Neptune's rings proved to be just as unusual as Uranus's. There are only four very narrow dark rings. Like Uranus's rings, they are probably made mostly of rocky dust. While the outer ring is

---

FAST FACTS ABOUT NEPTUNE

DIAMETER: 30,778 miles (49,530 km)—3.8 times larger than Earth

MASS: 17.14 times more than Earth

SURFACE GRAVITY (at equator): 1.14 times greater than Earth

LENGTH OF DAY: 16.1 Earth hours

LENGTH OF YEAR: 164 Earth years

DISTANCE FROM THE SUN: 2,798,972,020 miles (4,504,300,000 km)—30.1 times farther away than Earth is

Neptune's rings photographed by *Voyager 2*. The outer ring is unique in that it has strange clumps, probably caused by the gravitational effects of small, nearby moons. [NASA/JPL]

78,000 miles (126,000 km) wide, it is only 30 miles (50 km) thick. The outer ring, however, is unusual because it is not a uniformly smooth circle. Instead it has several broad, arclike concentrations of material. These are called **ring arcs**. They are about 15 miles (24 km) wide and 100 miles (161 km) long.

There are at least three ring arcs and maybe as many as a dozen. Something is causing the particles that compose the rings to bunch up in this strange way, but what that might be is unknown. *Voyager 2* discovered six tiny moons ranging from about 30 to 250 miles (50 to 400 km) within and just outside the ring system. Perhaps gravitational effects from these and other, yet-undiscovered moons are somehow creating the strange arcs.

CHAPTER SEVEN
MANY MOONS

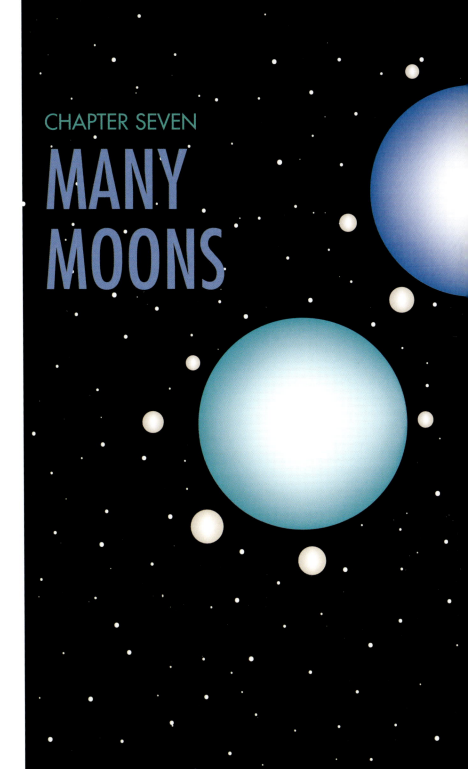

Before the *Voyager 2* flyby, Neptune was known to have at least two moons and Uranus five. One of Neptune's moons, Triton, is one of the largest moons in the solar system. Discovered in 1846, it is 1,682 miles (2,706 km) in diameter. It orbits 220,448 miles (354,760 km) from Neptune. From this distance, the planet would loom more than sixteen times larger than a full moon in Earth's sky.

Triton is an unusual moon. For one thing, it orbits Neptune *backward*. Seen from above their north pole, most of the solar system's large moons—such as Earth's—orbit in a counterclockwise direction. This direction is referred to as **prograde**. Triton's orbit, however, is clockwise, or **retrograde**.

Its orbit also has a very large tilt compared to the plane of Neptune's equator. Most of the other large moons in the solar system orbit near the plane of their planet's equator. The orbit of Earth's moon, for example, is tipped only 5.6 degrees. Triton's tilt of 23 degrees is therefore very unusual.

Astronomers think that the reason the other large moons orbit at or near the plane of their planet's equator is because they were formed around the same time as the planet, out of the same rotat-

(45)

ing cloud of dust and gas. This is why the moons have the same rotational direction as their planet. Triton's retrograde orbit and high orbital tilt suggest that it may have had a different origin.

Some scientists think that Triton might once have been an entirely independent planetary body that was later captured by Neptune's gravity. Supporting this idea, Triton seems to bear a strong resemblance to Pluto, the outermost planet in the solar system. Pluto is thought to be one of the large, icy bodies left over from the formation of the solar system. A vast cloud of these flying icebergs—called the **Oort cloud**—surrounds our solar system. There may have been many Pluto-sized objects in the outer regions of the solar system during the early days following its formation. One of these may have approached Neptune near enough to be captured into an orbit that—by sheer chance—happened to be retrograde.

In order for Neptune to have captured Triton, the moon must have been slowed down. Otherwise it would have looped past the planet and sailed back off into space again. Perhaps it approached so close to Neptune that the atmosphere of the planet slowed it down, in much the same way that friction from the atmosphere of Earth slows down a returning spacecraft. Or perhaps Triton smashed into one of Neptune's original moons. This collision would not only have slowed down Triton, the resulting debris would have provided material for Neptune's rings.

Triton's strange orbit is not the last of its unusual characteristics. In 1983, scientists studying infrared radiation from Triton found evidence that Triton possessed a thin atmosphere containing

The size of Neptune (blue) compared with Triton (brown). Earth and the Moon are at the lower left for comparison.

Triton [NASA/JPL]

MOONS AND PLANETS

How do astronomers distinguish between a moon and a planet? A planet is a solid (or partially liquid) body, such as Earth or Jupiter, orbiting a star. A planet also gives off no light of its own—it can be seen only by the light it reflects from its sun. A moon orbits a planet. A better word than moon is *satellite* (which means companion). Our Moon is the satellite of Earth, while Triton is a moon of Neptune and Oberon and Titania are moons of Uranus. Sometimes they are called natural satellites to distinguish them from artificial ones such as the Hubble Space Telescope or the International Space Station.

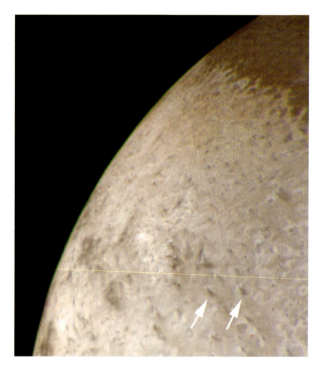

The bizarre surface of Triton: The small dark streaks in the middle of the image (two of which are indicated by arrows) may be deposits from geysers. [NASA/JPL]

Facing page: One of the powerful vents on Triton is sending a plume of gas many miles into the sky. Another vent is in the distance. The uppermost part of its plume has reached high-altitude winds in Triton's thin atmosphere, which have caused the plume to shear off at a right angle.

methane (because different compounds reflect infrared light in different ways). It is one of the only moons in the solar system known to have an atmosphere. Most moons are simply not massive enough to have sufficient gravity to retain gases and keep an atmosphere from escaping into space. Very large moons, such as Titan, the giant moon of Saturn, and Triton, however, are massive enough to retain the gases.

During the *Voyager 2* flyby of Neptune and Triton in 1989, the spacecraft's spectroscopes confirmed the existence of an atmosphere, along with the information that it is composed mostly of nitrogen, which is also the main component of Earth's atmosphere. Triton's atmosphere, however, is 100,000 times less dense than Earth's. This is very thin—hardly any atmosphere at all, in fact. It is as thin as Earth's atmosphere more than 50 miles (80 km) up, at the very fringe of space.

The existence of an atmosphere was not the last surprise. When *Voyager 2* flew by Triton, it discovered thin layers of haze 3 to 6 miles (5 to 10 km) above Triton's surface. Even more amazingly, it revealed huge geyserlike eruptions venting columns of dark vapor high into Triton's dark sky. These columns rise almost perfectly straight up for miles, until they meet high-altitude winds that shear them off at right angles into long streamers. Material from the geysers that has settled onto the surface of Triton has created long dark streaks across the landscape.

The photos of Triton taken by *Voyager 2* revealed that its landscape is also composed of flat plains of ice, meandering fissures (or cracks), and strange-looking bumpy areas dubbed "cantaloupe terrain." The surface is mostly methane and nitrogen ice. The poles

Triton's "cantaloupe" terrain may have been caused by melting ice. [NASA/JPL]

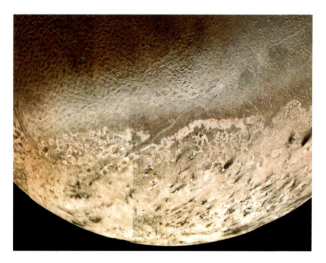

The south polar region of Triton [NASA/JPL]

are covered by seasonal caps of nitrogen frost. The surface temperature on Triton—thirty times farther away from the Sun than Earth—is a frigid –391°F (–235°C). The coldest temperature ever recorded on Earth was –128.6°F (–88.9°C) in Antarctica.

Triton's Tides

Triton's surface also appears to be young. The relative scarcity of craters implies that the surface has been periodically renewed, destroying old craters in the process. What might have caused this renewal? Triton's unusual history might hold the answer to this question. It may also hold the answer to the source of energy that powers Triton's huge geysers.

To understand what may be happening to Triton's surface, we need to go back in time to when Triton was first captured by Neptune. At first, the moon probably had a very elliptical orbit. As Triton went around Neptune, the orbit would bring Triton very close to the planet and then swing it far out. This would have resulted in unequal **tidal forces** that eventually caused the orbit to become more circular.

The unequal pull of one body on another one creates **tides**. If a moon is fairly close to its planet, the pull of gravity from the planet will be significantly greater on the side of the moon closest to the planet and much less on the side farther away. So in addition to making the orbit more circular, this unequal pull will also cause the moon to flex, much like a rubber ball squeezed over and over in your fist. If you try doing this yourself with a rubber ball, you will find the ball growing warmer. In the same way, the flexing of the moon will cause it to grow warmer.

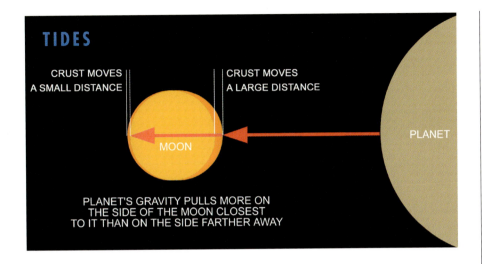

If the moon is close enough to its planet for the tidal effects to be strong, the interior of the moon may actually grow hot enough to melt. Forces like these would have heated Triton, creating a molten interior that would have stayed hot for a billion years. The volcanic eruptions that would have resulted from this might have periodically covered the surface with fresh lava flows. These would have erased old craters, which is why we see a young surface today.

Are the geysers we see now powered by this internal heat? If Triton was captured in the early years of the solar system's formation—say, 3 or 4 billion years ago—it would have been cool for a long time, perhaps too cool to have erupting geysers. Does the existence of the geysers mean that Triton was captured more recently, or is there another source of energy for the vents? No one knows for sure.

This image shows an area on the surface of Triton about 300 miles (483 km) wide. The smooth areas may be old impact craters that have been filled in with molten material. [NASA/JPL]

THE MOONS OF NEPTUNE

NAME	DATE OF DISCOVERY	DISTANCE FROM NEPTUNE IN MILES (KILOMETERS)	SIZE (RADIUS)* IN MILES (KILOMETERS)
Naiad	1989	14,579 (23,461)	19 (29) ?
Thalassa	1989	15,727 (25,309)	25 (40) ?
Despina	1989	17,250 (27,760)	47 (75)?
Galatea	1989	23,108 (37,187)	50 (80)?
Larissa	1989	30,313 (48,782)	65 x 55 (104 x 89)
Proteus	1989	57,716 (92,881)	135 x 125 (218 x 201)
Triton	1846	205,058 (329,994)	841 (1,353)
Nereid	1949	3,410,637 (5,488,634)	106 (170)

*Moons with two dimensions are not spherical. In this case, the longest and narrowest dimensions are given. The sizes with question marks are uncertain.

More Moons

Neptune has several other moons, all much smaller than Triton. One of the largest of these, Nereid, was discovered in 1949. It is only 211 miles (340 km) in diameter and orbits more than 3.4 million miles (5,488,634 km) from Neptune. From this distance Neptune would appear only about a quarter the size of a full

Neptune seen from its most distant moon, tiny Nereid

TOO SMALL TO BE ROUND

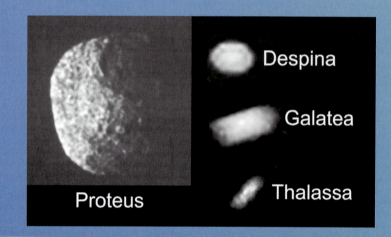

Four of Neptune's small, inner moons, photographed by *Voyager 2*. The largest, Proteus, is 135 by 125 miles (218 by 201 km), while the smallest, Thalassa, is only about 25 miles (40 km) wide. All of them are too small to be spherical. Very little is known about these moons. Even their sizes are uncertain. [NASA/JPL]

All of the planets in the solar system and most of the moons are spherical. But there are many moons (and most asteroids) that are not round at all. Jupiter's moon Amalthea is shaped something like a hockey puck, while the two tiny moons of Mars, Phobos and Deimos, look like potatoes. Large bodies, however, tend to be spherical because their gravity pulls the material from which they're made into that shape. The more material, or mass, there is, the stronger the gravity and the more spherical the object will become. But sometimes an object may be too small for its gravity to overcome the strength of the material it's made of. When this happens, the object can be almost any shape at all.

Facing page: A view of Neptune from its tiny innermost moon, Naiad

moon back on Earth. Nereid's orbit is tipped 30 degrees in relation to Neptune's equator, suggesting that it might have been captured by Neptune. Its composition closely resembles that of a distant icy asteroid named Chiron, which orbits between Saturn and Uranus. Since Chiron is thought to be a planetesimal left over from the formation of the solar system, astronomers suspect that this might be the origin of Nereid as well.

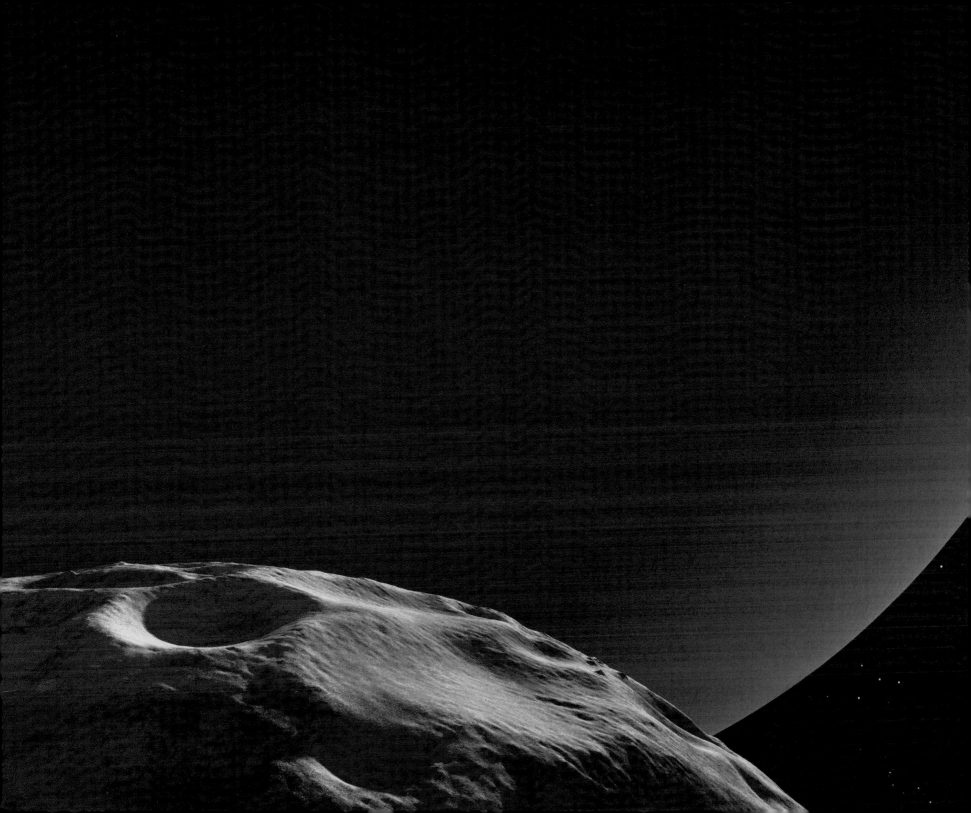

The other moons all orbit much closer to Neptune than Triton, most of them near or within the rings. These little moons were all discovered by the *Voyager 2* spacecraft. They are probably irregular lumps of ice, ranging in diameter from 20 to 130 miles (32 to 210 km), too small to have enough gravity to pull them into spheres.

Uranus's Strange Moons

Uranus has perhaps the largest collection of unusual moons in the solar system. There are twenty-one of them at last count. They are almost all named after characters in the plays and poems of William Shakespeare and Alexander Pope, although there's no special reason for this. The first of Uranus's moons to be discovered were named after Shakespeare's and Pope's characters, and astronomers thought it would be nice to continue doing so.

The five largest moons were all discovered before the *Voyager 2* flyby in 1986. They range in size from 293 miles (472 km) to 980 miles (1,578 km) in diameter. The very dark, small inner moons were discovered by *Voyager 2*, while the five most distant moons were all found recently by Earth-based observers. Unlike the moons of Neptune but like most of the other moons in the solar system, Uranus's moons orbit in the plane of the planet's equator.

The innermost of the five large moons, Miranda, was discovered in 1948. It is just 147 miles (236 km) in diameter. *Voyager 2* photos revealed a world that must have had an extraordinarily

THE MOONS OF URANUS

NAME	DATE OF DISCOVERY	DISTANCE FROM NEPTUNE IN MILES (KILOMETERS)	SIZE (RADIUS)* IN MILES (KILOMETERS)
Cordelia	1986	15,034 (24,193)	8 (13)?
Ophelia	1986	17,526 (28,204)	10 (16)?
Bianca	1986	20,883 (33,607)	14 (22)?
Cressida	1986	22, 500 (36,208)	21 (33)?
Desdemona	1986	23,053 (37,099)	18 (29)?
Juliet	1986	24,110 (38,799)	26 (42)?
Portia	1986	25,190 (40,538)	34 (55)?
Rosalind	1986	27, 570 (44,368)	18 (29)?
Belinda	1986	30,882 (49,697)	21 (34)?
1986U10	1999	47,226 (76,000)	25 (40)
Puck	1985	37,561 (60,445)	48 (77)
Miranda	1948	64,775 (104,241)	147 (236)
Ariel	1851	102,954 (165,681)	360 (579)
Umbriel	1851	149,410 (240,441)	364 (585)
Titania	1787	254,949 (410,281)	490 (790)
Oberon	1787	346,145 (557,041)	472 (760)
Caliban	1997	4,438,872 (7,143,341)	19 (30)?
Stephano	1999	6,214,000 (10,000,000)	12.4 (20)
Sycorax	1997	7,573,649 (12,188,041)	37 (60)?
Prospero	1999	10,253,000 (16,500,000)	12.4 (20)
Setebos	1999	10,986,973 (17,681,000)	12.4 (20)

*Moons with two dimensions are not spherical. In this case, the longest and narrowest dimensions are given. The sizes with question marks are uncertain.

In this view of Miranda, it's easy to see how its jumbled surface resembles a patchwork quilt. At the top is a huge cliff, called the Great Wall, that resulted when two blocks of Miranda's crust shifted. [NASA/JPL]

violent history. It looks very much as though it had been literally shattered into pieces and imperfectly reassembled. The landscape is a tortured collection of valleys, cracks, grooves, and steep cliffs. One of the latter is a glass-smooth wall over 3 miles (5 km) high with a slope of about 45 degrees. The cliff was formed when two vast blocks of Miranda's crust moved—one up, the other down.

What caused Miranda to nearly break into pieces? Its proximity to Uranus probably had something to do with it. The planet is only 80,782 miles (130,000 km) away. From that distance Miranda would be eighty-eight times larger in Uranus's sky than a full moon here on Earth. Uranus would probably cause terrific tidal stresses in Miranda. But that might not be the full reason for Miranda's shattered appearance.

A giant planet like Uranus attracts a great many meteoroids and asteroids. Since Miranda is so close to Uranus, it is in more jeopardy of running into one of these asteroids than an outer moon would be. One calculation determines that it is fourteen times more likely to be hit than an outer moon. It is possible that one of these asteroids was large enough to shatter Miranda into pieces. The pieces may have remained in orbit around Uranus, where they eventually reassembled. Some astronomers have even suggested that this may have happened several times. While this violent history would certainly explain Miranda's weird, piecemeal appearance, the theory has fallen into some disfavor, and most astronomers now think that the moon's tortured landscape can be explained entirely by tidal forces.

Miranda may have been shattered into pieces by a tremendous collision, which would explain its present-day piecemeal appearance.

A view across the grooved terrain of Miranda

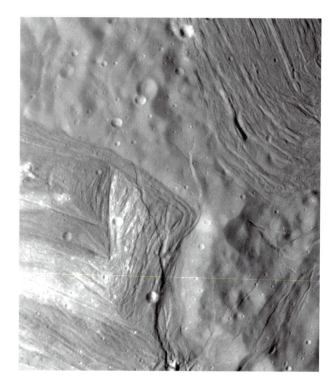

Above: A region on Miranda known as the *chevron* because of the V-shaped pattern of the grooves [NASA/JPL] Right: The Great Wall of Miranda is a fabulous, sheer-walled cliff 3 miles (5 km) high. [NASA/JPL]

Facing page: The Great Wall of Miranda

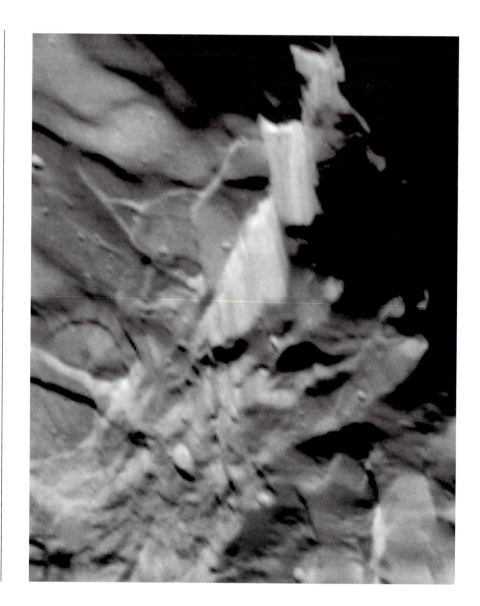

An overall view of Ariel taken by *Voyager 2*. Its surface is covered with deep, crisscrossing valleys and canyons. Smoother areas have probably been resurfaced by volcanic action. [NASA/JPL]

Ariel, at 360 miles (579 km) in diameter, is about twice the size of Miranda and is the next large moon out from Uranus. It was discovered in 1851. Its heavily cratered landscape is crisscrossed by broad, flat-bottomed canyons hundreds of miles long and up to 6 miles (10 km) deep. Like all of Uranus's large moons, it is 40 to 50 percent water ice, with the remainder rock. It may have been molten in the distant past but has long since been frozen. The canyons may have been created when the surface cracked as the moon froze.

Umbriel is the third most distant of the large moons. Like Ariel, it was also discovered in 1851 and is only slightly larger than Ariel. Also like Ariel, its surface is heavily cratered and very dark, probably because there is more **carbonaceous** dust and rocks mixed in with its ice.

Titania, at 980 miles (1,578 km) in diameter, is the largest of Uranus's satellites. It is the fourth most distant of the large moons from Uranus. It was discovered in 1787 by William Herschel, who also discovered Uranus. It orbits Uranus at a distance of 254,949 miles (410,281 km). From this distance Uranus would appear slightly more than twelve times larger than a full moon back on Earth. Titania's surface, which appears to be a mixture of ice and carbonaceous soil, is heavily cratered, with systems of interconnected valleys up to 930 miles (1,500 km) long and 60 miles (100 km) wide. They may have been created when the original molten surface froze.

Oberon, the fifth most distant large moon, is a near twin of Umbriel. Its heavily cratered surface is not particularly distinctive.

Some of the craters seem to have dark floors, which might be the result of dirty water oozing from beneath the surface and freezing. The exception to this otherwise ordinary landscape is at least one extraordinarily high mountain that looms between 3.7 and 6.8 miles (6 to 11 km) above the surrounding area. In comparison, Mount Everest, the highest mountain on Earth, is 5 miles (8 km) above sea level. This is impressive enough, but considered in relation to the size of Oberon, however, this mountain takes on a whole new scale. Oberon is more than eight times smaller than Earth. Its mountain covers 1/86th the diameter of Oberon, while Mount Everest covers only 1/1,585th the diameter of Earth. This makes Oberon's mountain proportionally eighteen times higher than Everest!

    The remainder of Uranus's moons, five of them farther out than Oberon and eleven closer than Miranda, are only relatively tiny lumps of ice and rock, the largest of which is only 100 miles (60 km) wide. Most astronomers believe that these moons are probably captured asteroids.

The fractured surface of Ariel has intersecting valleys like the ones shown here. In the distance is Uranus, with its dark rings seen edge-on, and Miranda, the next nearest moon.

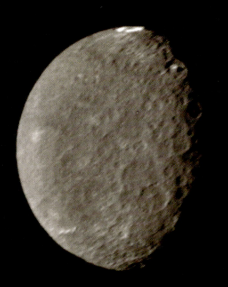

Voyager 2 reveals Umbriel's heavily cratered surface in this photo. The darkest of Uranus's moons, it features a prominent crater near the top right, with a bright central peak. The bright ring at the top edge of the moon is an unusual feature. It may be a frost-filled crater. [NASA/JPL]

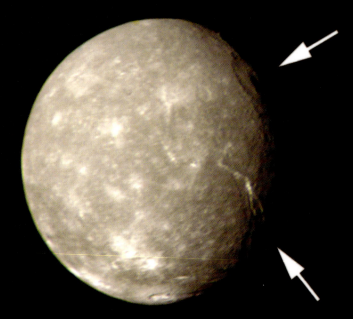

In this view of Titania a large canyon-like feature is visible at the right (lower arrow), as well as a large impact crater (upper arrow). [NASA/JPL]

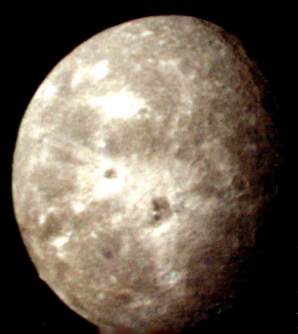

This view of Oberon was taken by *Voyager 2*. It shows several large impact craters. Near the center is a crater with a bright central peak and a dark floor. A large mountain about 4 miles (6 km) high can be seen at the lower left edge of the moon. [NASA/JPL]

This large crater on Umbriel has a central peak that rises miles above the surrounding terrain. The bottom of the crater is covered in ice. Uranus looms in the sky along with two of its other moons, Ariel (right) and Miranda (left). The rings are casting curved shadows on the planet.

This is Uranus seen from one of Titania's great valleys. Uranus's dark rings are just barely visible, as is Umbriel, the next nearest moon.

CHAPTER EIGHT
EXPLORING THE SOLAR SYSTEM

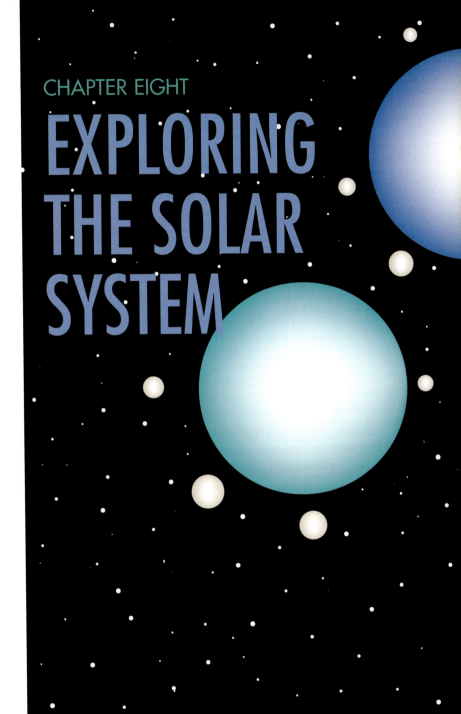

Human beings have explored only two bodies in the solar system: Earth and the Moon. All other planet and satellite exploration has been by robots—complex spacecraft that have sent back enormous amounts of information, from close-up photos to actual samples. Because of these instruments, we have learned almost as much about some of the planets as we know about Earth. Probes have made incredible discoveries: the rings of Jupiter and Neptune, the volcanoes of Io, the weird landscapes of Miranda, the geysers of Triton, and hundreds of others.

The first probe to reach another world was the Russian *Luna 2*, which crashed into the Moon in 1959. The American *Ranger 7* took the first close-up photos of the Moon in 1964, while the Russian *Luna 9* actually landed in 1966, as did the American *Surveyor 1* later the same year.

The first probe to successfully reach another planet was *Mariner 2*, which flew by Venus in 1962. *Mariner 4* made a flyby of Mars in 1964 and sent back the first close-up pictures of the planet. *Pioneer 10* made the first flyby of Jupiter in 1972, and *Pioneer 11* flew by both Jupiter and Saturn in 1973. There have been dozens of others, from the *Venera* spacecraft that landed on Venus to the recent *Pathfinder* that placed the Sojourner rover on Mars in 1997.

One of the most successful space probes ever launched was *Voyager 2*, which was launched from Earth in August 1977. It flew by Jupiter, Saturn, Uranus, and Neptune, returning mountains of data about each planet, including thousands of high-resolution photos of the planets and many of their moons.

Voyager 2 arrived at Uranus in 1986 and Neptune in 1989. After its encounter with Neptune, *Voyager 2* headed out of the solar system. At the end of 2001 it was 9,730,000,000 miles (15,665,300,000 km) from the Sun. We will probably lose all contact with the spacecraft by the year 2016. At that time it will be 100 AU from the Sun—100 times farther away from the Sun than Earth is and more than twice as far from the Sun as Pluto, the most distant planet in the solar system. In 358,000 years, *Voyager 2* will pass within 0.8 light-years of the star Sirius. On board is a golden record carrying the sounds of Earth and the music, pictures, and words of the people who live there—a greeting card to the universe.

Back to the Frontier?

Will spacecraft again be sent to the mysterious worlds at the frontier of the solar system? There are many mysteries remaining about Uranus and Neptune—even more than before *Voyager 2* made its brief visit. What happened to Miranda? Why does Uranus have such an unusual tilt? What causes the strange arcs in Neptune's rings? Why does Triton resemble Pluto so much? We won't know until we return.

The last view of Neptune (below) and Triton (top) as *Voyager 2* departed [NASA/JPL]

GLOSSARY

accretion: the process by which larger bodies are built up from smaller ones by mutual gravitational attraction.

astrology: a pseudoscience that suggests that human lives are influenced or controlled by the positions of planets and stars. This idea is rejected by modern astronomers and other scientists.

asteroid: a medium-sized rocky object orbiting the Sun; smaller than a planet, larger than a meteoroid. Also known as a *planetoid*.

AU (astronomical unit): the average distance of Earth from the Sun—92,960,000 miles (149,600,000 km).

axis (*plural* **axes**): the line around which a planet rotates.

carbonaceous: a substance that contains carbon or carbon compounds.

comet: a medium-sized icy object orbiting the Sun; smaller than a planet.

diffraction grating: a glass or plastic sheet engraved with thousands of fine grooves that break up light into a spectrum. (The rainbow colors you sometimes see on a CD are caused by the fine grooves in its surface acting as diffraction grating.)

elliptical: refers to something shaped like an ellipse, which is a flattened circle or oval.

equatorial plane: an imaginary plane that cuts through a planet at its equator.

gas giant: any large planet composed mostly of gas and liquid.

gravitational contraction: the effect of a body compressing under the influence of its own gravity. This can create great amounts of heat in the center of the object.

hydrocarbon: a chemical compound composed only of atoms of hydrogen and carbon. The simplest ones are gases at ordinary temperatures, while more complex ones are liquid or solid.

hydrogen: the lightest and most abundant element in the universe.

invert: to turn upside down.

molecule: two or more atoms attached together.

moon: any natural body orbiting a planet.

nebula (*plural* **nebulae**): a cloud of gas in space.

Oort cloud: a distant cloud of icy bodies that surrounds the solar system beyond the orbit of Pluto. Occasionally one of these icy bodies falls into the solar system as a comet. It was named for the Dutch astronomer, Jan Hendrik Oort, who first suggested its existence.

orbital period: the time taken for one body to circle another one. The orbital period of Earth around the Sun is one year, the orbital period of the Moon around Earth is one month.

planet: any large body orbiting a star.

planetesimal: any of the small, solid bodies that existed during the early stages of the solar system's formation.

prograde: orbiting or rotating in a counter-clockwise direction (as seen from above a north pole).

protoplanetary disk: a large disk of dust and gas surrounding a star that eventually accretes to form planets.

protostar: a cloud of dust and gas of stellar mass in an early stage of collapse. When it collapses enough, it will form a star.

pseudoscience: a false science that has no basis in fact.

retrograde: orbiting or rotating in a clockwise direction (as seen from above a north pole).

ring arc: an independent segment of a planet's ring.

satellite: any body orbiting another body. The Moon is a satellite of Earth, while Earth is a satellite of the Sun.

seasons: effects of the axial tilt of a planet. The tilt will cause any one region on a planet to receive more and less warmth from the Sun as it orbits it.

spectrograph: an instrument for recording the image of a spectrum.

spectroscope: an instrument that breaks down light into its spectrum.

spectrum: the band of colors created when light is broken up by a prism or diffraction grating.

star: a mass of gaseous material that is massive enough to start nuclear reactions in its central region.

sunspots: gigantic magnetic storms on the Sun. Because these regions are cooler than the surrounding area, they appear dark.

terrestrial planet: Any rocky/metallic planet; named for its resemblance to Earth.

tidal forces: the effect of an unequal gravitational pull of one body upon another. This unequal pull can create great stresses in the smaller body that can cause it to heat up. If the stresses are great enough, the body may be pulled apart.

tide: an uneven gravitational attraction between two objects.

For More Information

Web sites

Alpha Centauri's Universe
http://www.to-scorpio.com/index.htm
A good site for basic information about the solar system.

Fact Sheet: Neptune Science Summary
http://vraptor.jpl.nasa.gov/voyager/vgrnep_fs.html
An official NASA Web site with information from the Voyager mission to Neptune.

Fact Sheet: Uranus Science Summary
http://vraptor.jpl.nasa.gov/voyager/vgrur_fs.html
An official NASA Web site with information from the Voyager mission to Uranus.

NASA Spacelink
http://spacelink.msfc.nasa.gov/index.html
Gateway to many NASA Web sites about the Sun and planets.

Nine Planets
http://www.nineplanets.org
Detailed information about the Sun, the planets, and all the moons, including many photos and useful links to other Web sites.

Official Voyager Project Home Page
http://vraptor.jpl.nasa.gov/
Everything you want to know about the missions of *Voyager 1* and *2*.

Planet Orbits
http://www.alcyone.de
A free software program that allows the user to see the positions of all the planets in the solar system at one time.

Planet's Visibility
http://www.alcyone.de
A free software program that allows users to find out when they can see a particular planet and where to look for it in the sky.

Solar System Simulator
http://space.jpl.nasa.gov/
An amazing Web site that allows the visitor to travel to all the planets and moons and create their own views of these distant worlds.

Books

Beatty, J. Kelly, Carolyn Collins Petersen, and Andrew Chaikin, eds. *The New Solar System*. Cambridge, MA: Sky Publishing Corp, 1999.

Clay, Rebecca. *Space Travel and Exploration.* Brookfield, CT: Twenty-First Century Books, 1997.

Hartmann, William K. *Moons and Planets.* Belmont, CA: Wadsworth Publishing Co., 1999.

Miller, Ron, and William K. Hartmann. *The Grand Tour.* New York: Workman Publishing Co., 1993.

Schaaf, Fred. *Planetology.* Danbury, CT: Franklin Watts, 1996.

Spangenburg, Ray, and Kit Moser. *A Look at Moons.* Danbury, CT: Franklin Watts, 2000.

Vogt, Gregory. *Deep Space Astronomy.* Brookfield, CT: Twenty-First Century Books, 1999.

Magazines

Astronomy
http://www.astronomy.com

Sky & Telescope
http://www.skypub.com

Organizations

American Astronomical Society
2000 Florida Avenue NW, Suite 400
Washington, DC 20009-1231
http://www.AAS.org

Association of Lunar and Planetary Observers
P.O. Box 171302
Memphis, TN 38187-1302
http://www.lpl.arizona.edu/alpo/

Astronomical Society of the Pacific
390 Ashton Avenue
San Francisco, CA 94112
http://www.aspsky.org

The Planetary Society
65 N. Catalina Avenue
Pasadena, CA 91106
http://planetary.org

INDEX

Page numbers in *italics* refer to illustrations.

Accretion, 12
Acetylene, 43
Adams, John Couch, 18, 21
Airy, George, 18, 21
Amalthea, 54
Andromeda Nebula, 19
Ariel, 57, 64, *64*, *66–67*
Arrest, Heinrich d', 21
Asteroids, 20, 26, 58, 65
Astrology, 6, 17, 19
Axis, 25

Belinda, 57
Beta Pictoris, 12
Bianca, 57
Bode, Johann, 17, 20
Bode's Rule, 20

Caliban, 57
Callisto, 8
Cantaloupe terrain, 48, *50*
Carbonaceous dust, 64
Challis, James, 21

Chiron, 54
Comets, orbits of, 16
Cordelia, 36, 37, 57
Cressida, 57

Deimos, 54
Desdemona, 57
Despina, 52, 54
Diffraction grating, 24
Discourses on the Plurality of Worlds (Fontenelle), 9

Earth, 12, *13*, 14, 16
  atmosphere of, 48
  axis of, 25, 26
  distance from sun, 20
Egyptians, ancient, 6
Elliptical orbits, 16
Equatorial plane, *25*, 25
Ethane, 43
Europa, 8

Fontenelle, Bernard de, 9

Galatea, 52, 54
Galilean satellites, 8

Galileo Galilei, 7, 7–9
Galle, Johann, 21
Ganymede, 8
Gas giants, 12, *13*, 14
George III, King of England, 16
Glass prism, 24
Gravitational contraction, 11, 40
Great Dark Spot, 38, 40
Great Red Spot, 38
Great Wall of Miranda, *58*, *62*, *63*

Helium, 15, 23, 43
Herschel, Caroline, 19
Herschel, Sir William, 10, *16*, 16, 19, 64
Holberg, Ludwig, 9
Horoscopes, 6
Hubble Space Telescope, 11, 12, 35, 47
Hussey, T.J., 17–18
Hydrocarbons, 43
Hydrogen, 12, 15, 23, 43

International Space Station, 47
Io, 8, 71

Juliet, 57

(78)

Jupiter, 7, 9, *13*, 14, 40
    axis of, 25
    composition of, 22
    distance from sun, 20
    Great Red Spot, 38
    moons of, 8, 54
    ring system of, 34, 36, 71
    Voyager 2 flyby, 31, 32, 72

Kepler, Johannes, 9

Larissa, 52
Lassell, William, 21
Le Verrier, Urbain, 21
Locke, Richard Adams, 10
Luna 2, 71
Luna 9, 71

Mariner 2, 71
Mariner 4, 71
Mars, 7, *13*, 14
    distance from sun, 20
    exploration of, 71
    moons of, 54
Mercury, 7, 9, *13*, 14
    axis of, 25
    distance from sun, 20
Meteoroids, 58
Methane, 23, 24, 34, 43, 48
Miranda, 57, *59*
    discovery of, 56
    Great Wall of, *58*, *62*, *63*
    surface of, *58*, 58, *60*, *61*, 71
Moon, 6, 8–10, 71

Moons. (*see also* Miranda; Triton)
    of Jupiter, 8, 54
    of Mars, 54
    of Neptune, 5, 24, 52, 54, 56
    of Uranus, 5, 24, 26, 28, 36, 37, 45, 47,
        56–58, *64*, 64–65, *66–67*, *68*, *69*
Mount Everest, 65

Naiad, 52
Natural satellites, 47
Nebulae, 19
Neptune, *13*, 14, *53*, *55*
    atmosphere of, 40, 42, 43
    composition of, 22–23
    Dark Spot 2, *39*, 39
    discovery of, 5, 20, 21
    distance from Earth, 5
    fast facts about, 34
    Great Dark Spot, 38, 40
    interior of, *42*
    moons of, 5, 24, 52, 54, 56 (*see also* Triton)
    orbit of, 20, 24
    ring system of, 36, 43–44, *44*, 46, 71
    rotation of, 22
    Scooter, *39*, 39
    size of, 24, 38
    size of Sun from, 23, *23*
    temperature of, 38, 42
    Voyager 2 flyby, *4*, 6, 31, 38, 42–44, *44*, 48, 73
Nereid, 52, 54
Newton, Sir Isaac, 24
1986U10, 57
Nitrogen, 48

Nuclear reaction, 11

Oberon, 47, 57, 64–65, *68*
Oort cloud, 46
Ophelia, 36, 37, 57
Orbital period, 36
Orion Nebula, 12

Pathfinder, 71
Phobos, 54
Piazzi, Giuseppe, 20
Pioneer 10, 71
Pioneer 11, 71
Planetesimals, 12, 54
Planets. (*see also* Solar system; specific planets)
    axes of, 25
    naming of, 7
    orbits of, 16
Pluto, 5, 12, 14–15, 20, 26, 46, 72
Pope, Alexander, 56
Portia, 57
Prograde orbit, 45
Prospero, 57
Proteus, 52, 54
Protoplanetary disks, 11–12, *12*
Protostars, 11
Pseudoscience, 6
Puck, 57

Ranger 7, 71
Raspe, Rodolf Erich, 10
Retrograde orbit, 45, 46
Ring arcs, 44
Ring system

of Jupiter, 34, 36, 71
of Neptune, 36, 43–44, *44*, 46, 71
of Saturn, 30, 34–36
of Uranus, 28, 30, *30*, 34–37, *35*, *36*
Rosalind, 57

Satellites, 47
Saturn, 7, 9, *13*, 14
   composition of, 22
   distance from sun, 20
   ring system of, 30, 34–36
   Voyager 2 flyby, 31, 32, 72
Seasons, 6
Setebos, 57
Shakespeare, William, 56
Shepherd moons, 36, 37
Solar system
   discovery of, 6–10
   evolution of, 11–15
   exploring, 71, 73
Somnium (Kepler), 9
Spectrograph, 24
Spectroscope, 22, 28, 48
Spectrum, 24
Stars, 6, 8, 11

Stephano, 57
Sun, 16, 20, *23*, 23, 34
Sunspots, 19
Surveyor 1, 71
Sycorax, 57

Telescope
   Herschel's, *19*, 19
   invention of, 7, 8
Terrestrial planets, 12, *13*, 14
Thalassa, 52, 54
Titan, 48
Titania, 47, 57, 64, *68*
Titius, 20
Triton, *46*, 47, 52, 73
   atmosphere of, 46, 48
   geysers of, 48, *49*, 50, 51, 71
   orbit of, 45–46
   surface of, 48, *48*, 50, 51
   tides, 50–51

Umbriel, 57, 64, *68*, *69*
Urania, goddess of astronomy, 17
Uranus, *13*, 14, *70*
   atmosphere of, 34

axis of, 26, 28
collision theory, 26, 28, *29*
composition of, 22–24
discovery of, 5, 16–17, 19
distance from Earth, 5
distance from Sun, 34
fast facts about, 34
interior of, *35*
moons of, 5, 24, 26, 28, 36, 37, 45, 47, 56–58, *64*, 64–65, *66–67*, 68, *69* (see also Miranda)
orbit of, 17–18, *18*, 23–24, *27*
ring system of, 28, 30, *30*, 34–37, *35*, *36*
rotation of, 22
size of, 24, 34
size of Sun from, *23*, 23
temperature of, 34
Voyager 2 flyby, *4*, 6, 31, 32, *33*, 34

Venera spacecraft, 71
Venus, 7, *13*, 14, 20, 71
Voyager 2, 4, 6, 31–34, *32*, 38, 42–44, *44*, 56, 64, 68, 72, 73

Water vapor, 40